Cambridge Elements ≡

Elements of Paleontology
edited by
Colin D. Sumrall, University of Tennessee

UNDERSTANDING THE TRIPARTITE APPROACH TO BAYESIAN DIVERGENCE TIME ESTIMATION

Rachel C. M. Warnock
Friedrich-Alexander Universität Erlangen-Nürnberg

April M. Wright
Southeastern Louisiana University

Paleontological
S O C I E T Y

CAMBRIDGE
UNIVERSITY PRESS

CAMBRIDGE
UNIVERSITY PRESS

University Printing House, Cambridge CB2 8BS, United Kingdom

One Liberty Plaza, 20th Floor, New York, NY 10006, USA

477 Williamstown Road, Port Melbourne, VIC 3207, Australia

314–321, 3rd Floor, Plot 3, Splendor Forum, Jasola District Centre,
New Delhi – 110025, India

79 Anson Road, #06–04/06, Singapore 079906

Cambridge University Press is part of the University of Cambridge.

It furthers the University's mission by disseminating knowledge in the pursuit of
education, learning, and research at the highest international levels of excellence.

www.cambridge.org
Information on this title: www.cambridge.org/9781108949422
DOI: 10.1017/9781108954365

First published 2020

A catalogue record for this publication is available from the British Library.

ISBN 978-1-108-94942-2 Paperback
ISSN 2517-780X (online)
ISSN 2517-7796 (print)

Understanding the Tripartite Approach to Bayesian Divergence Time Estimation

Elements of Paleontology

DOI: 10.1017/9781108954365
First published online: November 2020

Rachel C. M. Warnock
Friedrich-Alexander Universität Erlangen-Nürnberg
April M. Wright
Southeastern Louisiana University
Author for correspondence: Rachel C. M. Warnock,
rachel.warnock@bsse.ethz.ch

Abstract: Placing evolutionary events in the context of geological time is a fundamental goal in paleobiology and macroevolution. In this Element we describe the tripartite model used for Bayesian estimation of time-calibrated phylogenetic trees. The model can be readily separated into its component models: the substitution model, the clock model, and the tree model. We provide an overview of the most widely used models for each component and highlight the advantages of implementing the tripartite model within a Bayesian framework.

Keywords: Bayesian phylogenetics, paleobiology, species divergence times, morphological clock, fossil calibration

ISBNs: 9781108949422 (PB), 9781108954365 (OC)
ISSNs: 2517-780X (online), 2517-7796 (print)

Contents

1 Introduction

Phylogenetic inference, also known as phylogenetic tree inference or simply tree inference, is common in all facets of biology and estimating a phylogeny is a critical step in many comparative analyses. The fact that tree inference is common can obscure the underlying complexity of the task. When a researcher estimates a phylogeny, they are attempting to reconstruct evolutionary events that potentially occurred millions of years ago. In modern phylogenetics, inferring trees is often achieved by using an evolutionary model that ideally captures the generating processes that underlie our data. Since no two datasets are exactly the same, in terms of evolutionary history or taxon sampling, choosing the best approach to building a phylogeny requires deep knowledge of the taxonomic group, as well as phylogenetic theory. In this review, we focus on the models commonly used to infer phylogenies in macroevolution and paleobiology research.

The primary source of evidence used to infer evolutionary relationships are phylogenetic characters: molecular sequences and morphology in the case of living species, or morphology in the case of most fossils. The number of differences observed between the species included in the phylogeny are used to measure evolutionary distances and to group them together in the tree. This task, however, becomes more challenging if we also need to date the inferred phylogenies. This is because phylogenetic characters only contain information about how closely or distantly different species are related, that is, their *relative* evolutionary distance. Additional temporal evidence, which can be given by geological events or the fossil record, is required to calibrate trees to geological time (sometimes called *absolute* time). Otherwise, it is not straightforward to distinguish between rapid evolutionary rates over short intervals versus slow evolutionary rates over long intervals. Fig. 1 provides a recap of the most important features of a phylogeny (tips, nodes, and branches, which together comprise the tree topology) and shows an example of undated and dated phylogenetic trees. An undated phylogeny will typically have branch lengths in units that reflect the overall number of molecular or morphological character changes that have occurred between the two edges of a branch (i.e., between the ancestral node and the younger node represented at the edges of this branch, respectively), while a dated tree will use units of calendar time, such as years or millions of years.

Inferring time-calibrated trees is often achieved by jointly estimating the topology and node ages. In performing this analysis, researchers usually assume a tripartite model of evolution: one model that describes the accumulation of differences in character data, a second that describes the distribution

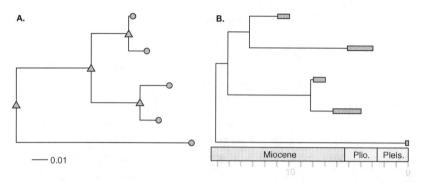

Figure 1 Undated versus dated phylogenetic inference. A phylogenetic tree is comprised of *tips* (indicated with circles in panel A), which represent the taxa between which we aim to infer the evolutionary relationships. These taxa are connected by *branches.* The branches are connected by *nodes* (indicated with triangles in panel A), which reflect the most recent common ancestor between two given tips. The overall structure of the tree used to represent phylogenetic relationships is referred to as the *topology.* In an undated phylogeny, branch lengths are typically in units that represent the overall amount of character change, indicated here by the scale bar. In undated model-based tree inference, the units usually represent the number of expected changes per character. A tree estimated with no temporal information can be seen in panel A. In a time-calibrated tree the branch lengths will be in units of calendar time, often in years or millions of years. Panel B shows the same tree as in panel A, but with branches in millions of years, along with stratigraphic ranges, i.e., the interval between first and last appearance times (gray boxes).

of evolutionary rates across the tree, and a third model describing the distribution of speciation events (node ages) across the tree (Drummond et al., 2006; Kishino, Thorne, & Bruno, 2001; Thorne, Kishino, & Painter, 1998; Yang & Rannala, 2006). This tripartite approach is a product of the history of phylogenetic model development and reflects the way in which researchers have traditionally tried to break down the problem of estimating evolutionary time from phylogenetic character data. While the tripartite model may sound complex, it enables researchers to treat each component as a discrete inferential module and provides them with the flexibility to combine different models that best suit their data. Understanding the tripartite framework is therefore useful for tracing the history of model development, making informed parameter choices, interpreting your results, and diagnosing problems with your analysis.

Within macroevolution and evolutionary biology, the tripartite framework has typically been applied to infer dated trees representing relationships among living species. Recent technical and theoretical advances have allowed the tripartite framework to be applied to trees that also include extinct representatives, that is, fossil species (Gavryushkina et al., 2017; Gavryushkina et al., 2014; Heath, Huelsenbeck, & Stadler, 2014; Ronquist et al., 2012; Stadler, 2010; Zhang et al., 2016). This means the framework can be applied to entirely extinct clades, or other datasets for which we rely on morphology, rather than molecular data, to inform phylogenetic inference. The tripartite model has been used to infer trees, times, and evolutionary rates among Cenozoic canids (Slater, 2015), crown birds and their Mesozoic relatives (Lee et al., 2014), Paleozoic echinoderms (Wright, 2017; Wright & Toom, 2017), Cambrian trilobites (Paterson, Edgecombe, & Lee, 2019), and Cambrian Cinctans (see Wagner, Wright, & Wright in this Element series).

We describe the components of the tripartite model in more detail and the processes they aim to capture. We then explain how to perform this analysis with Bayesian methods and highlight some of the advantages of using this statistical framework. Finally, we discuss how different aspects of the tripartite model can be linked and how this can be used to test hypotheses in paleobiology.

2 A Brief Introduction to Bayesian Inference in Phylogenetics

In this Element, we focus on divergence time estimation using Bayesian methods, which incorporate prior information and researcher intuition about parameters in our model. Unlike some other approaches, Bayesian methods estimate a sample of phylogenetic trees as well as a sample of values for the parameters of the underlying phylogenetic model. We can think of Bayesian inference as having three important components: the model likelihood, the prior, and the posterior.

We will be discussing these methods in a Bayesian context for a few reasons. Perhaps the most important is that Bayesian methods estimate a sample of plausible parameter values under a model. A Bayesian method inherently provides an indication of the uncertainty associated with any inferred model parameter. Since we are unlikely to be able to observe the true parameter values for an event that occurred millions of years in the past, it is prudent to consider possible ranges for parameters in our model, within which the true parameter is likely to be. Bayesian methods also allow researchers to constrain the values an individual parameter can take. This is a desirable property because we may have prior information from studies conducted by other researchers about the numerical value of a parameter. In this way, Bayesian inference provides an intuitive approach to accommodating uncertainty in other evolutionary and sampling parameters, and incorporating our existing knowledge of parameter values. On a practical level, much of the widely used divergence time-estimation software has also been written in a Bayesian context. See Box 3, "Maximum Likelihood and Bayesian Estimation," for a discussion of software for Bayesian analysis.

2.1 The Model Likelihood

We often think of statistical words in colloquial terms. For instance, we may think of "likelihood" in our daily life as referring to an event being likely or unlikely. This is different from statistics, when we calculate the model likelihood, or probability, of the observed data given a particular model. A model is a mathematical description of a phenomenon. Models are made up of parameters, which are thought to represent key factors of that phenomenon. The relationship between parameters is described through mathematical expressions. Many parameters of a model are treated as *random variables*. A random variable has an unknown value, for which candidate values will be tested as the inference of the given parameters proceeds. Bayesian analyses typically sample large numbers of solutions that explain how the data may have been generated under the specified model, with each sample appearing in proportion to its probability. In the tripartite model for divergence time estimation, all parts of the model (substitution model, clock model, and tree model) will be represented in the likelihood, as well as in the prior (Fig. 2).

2.2 The Prior

A prior specifies a probability distribution from which the value of a particular parameter may be drawn. Importantly, the true value of a parameter can fall outside the 95 percent limits of the prior distribution and still be estimated correctly. Priors can be enforced with varying degrees of strength. If the data strongly support a value for a parameter that is in conflict with the prior specified, that value can still be supported if the prior is not strongly enforced. Priors can also be chosen to offer maximal flexibility in the potential values for the parameter. For example, a weak intuition about the value of a parameter can be incorporated via a vague prior. In biology, it is fairly common to use distributions such as the Gamma or Exponential as priors, which can be very flexible depending on the centrality and/or shape parameters. An example of the flexibility that can be achieved using alternative priors is shown in Fig. 4, in the context of the clock model. Because reliable information with which researchers can inform prior choice is often unavailable, this flexibility may be considered desirable (Brandley et al., 2006).

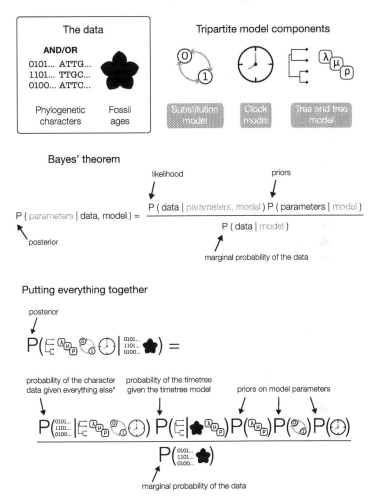

Figure 2 A tripartite model for Bayesian divergence time estimation. The top panel shows the key ingredients required during inference: The data used to generate time-calibrated trees, molecular or morphological phylogenetic characters, and age information, typically fossil sampling times. The model includes the substitution (site) model, which describes the evolution of characters, the clock model, which describes the distribution of evolutionary rates across the tree, and the tree model, which describes the distribution of speciation events across the tree. Bayes' theorem is presented in the middle panel. The bottom panel illustrates how everything comes together for the Bayesian estimation of divergence times. This figure is based on Fig. 1 in du Plessis and Stadler (2015). Note this formulation of Bayes' theorem treats fossil ages as fixed. In reality, fossils are associated with a range of ages and this uncertainty should be reflected through the use of priors. See Drummond and Stadler (2016) and Barido-Sottani, Aguirre-Fernández et al. (2019) for details.

Sometimes, the distinction between what we call a model and the prior can be difficult to discern (see Box 1, "The Likelihood, the Prior, and the Posterior" for more information). By constraining the values a parameter can take, it is possible to steer estimation toward or away from certain sets of values, without changing what facets of the generating process are being modeled. The priors, therefore, are part of a model, as they can lead to the parameters of that model taking on different values.

2.3 The Posterior

The posterior is the outcome of a Bayesian analysis and includes a distribution of plausible values for all of the parameters specified in our models, including the tree topology and divergence times. This component effectively combines the information from the prior with the likelihood. Fig. 2 provides a visual guide of how the prior and the likelihood come together to obtain the posterior. We cannot easily compute the posterior probability due to the relative complexity of phylogenetic models. The posterior is typically generated through what is referred to as Markov Chain Monte Carlo (MCMC) sampling. Under this algorithm, values for parameters are proposed, and the likelihood of the data under this model is scored. Then, the model parameters are changed, and the data are rescored under this new model. Generally, if the new parameter is an improvement, it is kept, and used as the seed for the next set of changes. MCMC does not track what values have already been scored, therefore a parameter that is a good fit may be returned to multiple times.

Values for each parameter will appear in the posterior distribution in proportion to how probable they are, given our model and priors. The highest point or points of the distribution represent the most probable parameter estimates. The variance of the posterior distribution for a given parameter reflects the uncertainty in that estimate. If the variance in our posterior distribution is relatively high, this reflects lots of uncertainty in our parameter estimate. Conversely, if the variance is relatively low the uncertainty in our estimate is low. Note that low uncertainty does not necessarily mean a value is true, just that there is high support for it given the data and model. In other words, even though the precision with which the parameter value has been estimated is high (low uncertainty, small variance), this does not imply the inference has also been accurate. A value could be estimated with low uncertainty but actually be very far from the true value. High precision does not guarantee high accuracy. Similarly, high variance or uncertainty in the posterior does not necessarily mean that the model is incorrect or that the analysis is bad. It simply means that there is limited information in our data. It is also possible to have more than one peak in your posterior sample. This indicates that multiple solutions are feasible given the model and the data.

From the posterior distribution of many standard model parameters (e.g., rate parameters) we can construct credible intervals, which are the Bayesian analog to confidence intervals. We typically use the 95% highest posterior density (HPD) interval, which is the spread of posterior values that contains 95% of the posterior. The upper and lower limits of the 95% HPD are an intuitive way of communicating the uncertainty associated with parameters such as rates or node ages. Summarizing a posterior distribution of trees, however, is altogether more tricky (Heled & Bouckaert, 2013; O'Reilly & Donoghue, 2018). There are a variety of strategies for capturing the phylogenetic relationships that are best supported by the posterior, which aim to summarize the estimated trees. Support values for each node are typically based on the proportion of trees in the posterior in which that node also appears. This is referred to as the posterior probability.

All approaches to producing summaries of the posterior have benefits and downsides (Heled & Bouckaert, 2013), especially when there is high uncertainty associated with the tree topology (O'Reilly & Donoghue, 2018). We emphasize that the posterior of a Bayesian phylogenetic tree inference is really a distribution of trees and associated model parameters. We should be careful to avoid placing too much confidence in any statistic or summary value from that distribution (Warnock, Yang, & Donoghue, 2017). Instead, it is important to understand the underlying models used to generate your tree, and how these may result in uncertainty given the data you have.

For more practical information about Bayesian phylogenetic inference we recommend Nascimento, dos Reis, and Yang (2017).

3 A Tripartite Model for Divergence Time Estimation

A model provides us with an expression for calculating the probability of observing our data, given some underlying assumptions about the processes that generated the data. Perhaps the most obvious thing we need to describe is the process of phylogenetic character evolution. This is achieved by using the *substitution model*, which describes the probability of changing between different character states. Substitution models are at the core of undated phylogenetic inference and essentially provide a measure of evolutionary distance. We tend to measure evolutionary distance or branch lengths in an undated phylogeny as the *number of expected substitutions per character*. We say "expected" substitutions because models allow for hidden state changes, such that the number of changes could be larger than the number we observe from our data. As noted above, phylogenetic characters do not contain information about absolute time. Evolutionary distances estimated using the substitution model actually represent a product of rate and time. Ultimately, for a dated phylogeny, we need to be able to estimate the substitution rate in *expected substitutions per character per calendar unit time.*

To extract information about rates and times from phylogenetic character data we need a modeling framework that describes the relationship between these variables, in addition to the substitution model. This is achieved with the addition of the two key model components required to date a phylogeny: the *clock model* and the *tree model.* The clock model describes how the substitution rate varies (or does not) across the tree. The tree model describes the process of speciation, extinction, and lineage sampling that generated the tree. To tease apart rate and time, we either need to know the average substitution rate or we need to calibrate the substitution rate using temporal information from elsewhere. For macroevolutionary timescales, calibration information typically comes from fossil-sampling times or the age of biogeographic events, information incorporated into the tree model. The tripartite approach to divergence time estimation is a hierarchical Bayesian model, which means it links together different submodels (i.e., the substitution, clock, and tree models); see Box 2, "Hierarchical Models" and Fig. 2.

Note that rate and time are often semi-identifiable, meaning that multiple combinations of parameters can potentially generate the same probability of the observed data. In this case, we may be unable to identify, or distinguish, the true parameter values. In practice, this means we need to put strong prior information on the average substitution rate or speciation times (dos Reis, Donoghue, & Yang, 2016; dos Reis & Yang, 2013). Consequently, the results will be very sensitive to these priors, so it is very important for biologists and paleobiologists to understand each of the component pieces in order to make good parameter choices.

4 Substitution Models

The first component of the tripartite model is the substitution model. The substitution model, sometimes called the site model, describes how phylogenetic characters in the dataset evolve. These models are called substitution models because they were initially written to describe nucleotide changes. These models describe how character change accumulates over time, leading to the observed phylogenetic data. In the context of divergence time estimation, phylogenetic data are typically either molecular or morphological data, although analyses that integrate both types of data have also been conducted

(Gavryushkina et al., 2017; Ronquist et al., 2012; Schrago, Mello, & Soares, 2013; Wood et al., 2013). While molecular and morphological data have very different properties. similar methods have historically been used to infer phylogenies from them.

Most data used in phylogenetic estimation has been discrete data. Discrete data can be broken into nonoverlapping categories. For example, nucleotide sequence data can be clearly separated into four states: adenine, cytosine, guanine, and thymine. Morphological characters are often divided into discrete states (de Queiroz, 1985). Most simply, these may correspond to an absence state (usually coded as "0") and a presence state (usually coded as "1") (Watrous & Wheeler, 1981). They may also correspond to more complex character diagnoses.

There are many models to describe how molecular sequence data evolve over time (Felsenstein, 1981; Hasegawa, Kishino, & Yano, 1985; Jukes & Cantor, 1969; Kimura, 1980; Tavaré, 1986). Nucleotide data tend to have well-defined and discrete properties. This allows a range of assumptions to be made about what changes we are likely to see over evolutionary time. In most common nucleotide substitution models, the probability of observing a change from one character state to another is taken to be the product of the *exchangeability* between two nucleotides at *equilibrium frequency* of the starting nucleotide (i.e., the nucleotide that exists in the sequence, which will be substituted for the other). The exchangeabilities refer to the probability of seeing a change from one particular state to another. These are often based on biochemical features of the nucleotide base. For example, it is unlikely to see a purine (two-ringed nucleotides, adenine and guanine) substituted for a pyrimidine (one-ringed bases, cytosine and thymine). This is due to biochemical properties – we are less likely to observe large changes, such as gaining a second ring of carbons on a structure, than smaller ones. Equilibrium frequencies refer to the frequency that we would see each of our character states if we allowed the evolutionary process to run infinitely (i.e., until it reaches the equilibrium). This is based on simple statistics: even if it is easy to change from one nucleotide to another, if the starting nucleotide is rare, that change will be seldom observed. It may be easy to transition from an adenine to a guanine but, if we have no adenines in our dataset, we are unlikely to observe this change over time.

Making different combinations of assumptions has yielded a panoply of molecular models. The simplest model of sequence evolution, the Jukes–Cantor model (Jukes & Cantor, 1969), assumes only one parameter: the rate of evolution. The exchangeabilities of this model are equal between all states. The equilibrium frequencies are also assumed to be equal. Therefore, under this model, you are as likely to observe a change that adds a second carbon ring to a pyrimidine as you are to observe changes from pyrimidines to other pyrimidines. On the opposite end of the spectrum, the general time reversible model (GTR) (Tavaré, 1986) allows for six different exchangeabilities, and for each molecular character to have its own equilibrium frequency, as illustrated in Fig. 3. This is a more complex model, but it is often supported as being the correct one for many datasets (Abadi et al., 2019). Molecular characters are typically assumed to evolve approximately neutrally, which means we can use relatively straightforward models of evolution.

Bayesian phylogenetics using morphological characters have historically used a more restricted set of models than analyses of molecular data. While we may be able to divide a discrete morphological character into multiple states, we may not be able to easily describe how states can transition from one to another over evolutionary time. For instance, molecular models assume that the biochemical properties of an adenine are the same today as they were in the past, and that all adenines are the same in different locations in the dataset. What are the properties of an absent morphological character? Does a change from state "0" to "1" at character "1" imply the same magnitude of changes as the same change of states at character "5"? The lack of consistent meanings to character states has limited the assumptions that can be made about the process that generated morphological data. Due to the limited number of morphological models available, model testing has not become common in morphological phylogenetics yet (though see an example of empirical model fitting in Bapst, Schreiber, & Carlson, 2017, and in Wagner et al. in this Element series), and understanding the role of the morphological model in divergence time estimation is an active area of scholarship (Klopfstein et al., 2019).

Because of the lack of common meanings between morphological character states, those working with morphological characters have largely been confined to working with the Mk model (Lewis, 2001) for discrete character evolution. This model is a translation of the Jukes–Cantor model (Jukes & Cantor, 1969) of sequence evolution to morphological characters, also shown in Fig. 3. Therefore, it makes the same assumptions about the generating process: that exchangeabilities are the same among all character states, and that all states have equal equilibrium character frequencies. This is a fairly restrictive model but, in a Bayesian context, some assumptions can be relaxed, allowing the user to make a variety of assumptions about the evolution of morphological data (Nylander et al., 2004; Wright, Lloyd, & Hillis, 2016). For a more detailed review of these methods, see Wright (2019). In addition, continuous morphological characters have recently been introduced in phylogenetic inference

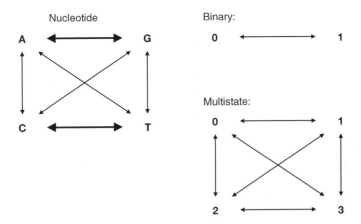

Figure 3 Phylogenetic Q matrices. In this schematic, we have several representations of different types of character change. For nucleotide data, we know that we are more likely to see certain types of change, such as two-ringed bases (purines) transitioning to other two-ringed bases, and one-ringed bases transitioning to other one-ringed bases. This is represented by thicker arrows connecting these bases. On the other hand, for morphological data, character states do not carry common meaning across characters. At one character, changing, for example, from a "0" state to a "1" state may be a small change. At another, it may mean gaining a complex character. Therefore, researchers have largely used the Mk model of Lewis (2001) to model these data. The schematic below shows the assumption of equal change probability between states.

(Goloboff, Mattoni, & Quinteros, 2006; Parins-Fukuchi, 2017) and divergence time estimation (Álvarez Carretero et al., 2019). The evolution of these continuous characters can be modeled under processes like Brownian motion (Felsenstein, 1973, 1985; Gingerich, 1993), or Ornstein–Uhlenbeck (Beaulieu et al., 2012; Butler & King, 2004; Hansen, 1997) or Lévy processes (Landis, Schraiber, & Liang, 2013), which allow for changes to accumulate continuously along a branch.

Discrete models are often adapted to take into account that characters (nucleotide or morphological) will evolve at different rates. Following Yang (1994), most researchers have modeled among-character rate variation (ACRV) as being distributed according to a Gamma distribution. A Gamma distribution can be manipulated to take a wide range of shapes. This distribution is then discretized into different categories (commonly four, but more categories are possible) and the median rate of each category is used as the rate of evolution for that category. This allows different sites to evolve according to different evolutionary rates, thereby correcting for different rates across sites. This practice is common for both molecular and morphological data, though some studies have indicated that lognormal-distributed ACRV may be more appropriate for morphology (Harrison & Larsson, 2015; Wagner, 2011). In particular, nonvariable or parsimony-uninformative characters are usually not collected by morphologists and the lognormal distribution potentially provides a better fit for datasets that do not include a zero-rate category. Not including these invariant characters is known to inflate rates of character change along branches, and must be corrected for in phylogenetic analysis (Leaché et al., 2015; Lewis, 2001).

5 Clock Models

Both the clock and tree models are required to tease apart rate and time, as well as to transform branches in units of time. The function of the clock model is to describe the way the rate of character change varies, or does not vary, across the tree. Individual models make different assumptions about how rate variation is distributed among branches. These range from every branch having the same rate of evolution to every branch having its own rate. Each of these models implies specific evolutionary dynamics. Below, we review some common clock models, which can apply to molecular or morphological data.

5.1 Strict Clock

Under the strict (or global) clock model, we assume that the rate of character change is constant across time and that the same rate applies to all branches in the tree (Zuckerkandl & Pauling, 1962, 1965). This model adds one parameter to the overall model, describing the conversion between the rate of character change and absolute time. Different values for this conversion are typically still sampled via MCMC in Bayesian analysis.

5.2 Uncorrelated Clock

Most clades, however, do have variation in the rate of evolution over time. A wide variety of clock models have been developed to describe how this variation manifests. One common family of clock models is the uncorrelated relaxed clock model. "Relaxed" refers to the clocks not being strict: any model that is relaxed will allow rate variation across the tree (Drummond et al., 2006; Drummond & Rambaut, 2007). "Uncorrelated" means that the rate of evolution on a particular branch is not dependent on the rates of evolution of its neighbors or ancestor. In this family of models, rates are typically assumed to be drawn from some distribution; the uncorrelated lognormal clock (UCLN) model being the most commonly used. Under this model, the rate of any particular branch is assumed to be drawn from a lognormal distribution independently of other branches (see an example in Fig. 4). The lognormal is a popular distribution in this type of analysis, as it implies most branches will have low, but typically nonzero, rates of evolution. Each branch has an independent draw from this distribution, meaning that the rate of a particular branch may be very different from its neighbors. The parameters of the lognormal distribution can be fixed, or can be estimated themselves (i.e., are hyperparameters). Nevertheless, other distributions, such as the exponential distribution, can also be used in these types of uncorrelated clock analyses. An exponential distribution, as seen in Fig. 4, implies some branch rates are very close to zero.

5.3 Autocorrelated Clock

The idea of a lineage's rate of evolution being independent of its ancestor's rate may strike some as odd. Much of the literature on clock models is focused on molecular data and molecular clocks. Molecular clocks are influenced by a variety of factors, such as generation times, population sizes, and metabolic rates (Bromham et al., 2015; Bromham, Rambaut, & Harvey, 1996; Gaut et al., 1992; Thomas et al., 2006). Morphological clocks are potentially impacted by the same variables, as well as other factors, such as developmental constraint (King et al., 2016). It would be reasonable, then, to expect that close relatives have similar evolutionary rates if they share these traits.

In autocorrelated rate models, the rate of a descendent branch is drawn from a probability distribution (Aris-Brosou & Yang, 2002) centered on the rate of the ancestor's branch. Different distributions can be assumed to allow the descendent's rate to be more different, or to force it to be more similar.

Autocorrelated clock models can also be continuous. A continuous autocorrelated clock model assumes that, again, the distribution from which the rate of a descendent is drawn is centered on the rate of evolution of the ancestor. Under these models, however, the variance is typically proportional to the length of the branch. More sophisticated assumptions can be made under these continuous autocorrelated relaxed clock models, such as the variance in rates evolving across the tree (Aris-Brosou & Yang, 2002, 2003; Kishino et al., 2001; Thorne et al., 1998; Thorne & Kishino, 2002).

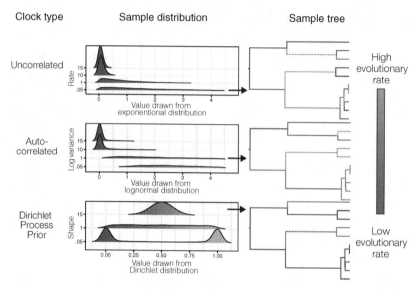

Figure 4 A schematic showing different clock models, and what they mean for the distribution of evolutionary rates across the tree. For each clock type, a set of sample distributions are shown. These distributions demonstrate how the distribution will look if a different prior is selected for its underlying parameters. An arrow indicates which distribution was used to simulate the rates shown on the sample tree. Row 1 shows an uncorrelated clock, with branch rates drawn from the exponential distribution. Because this clock is uncorrelated, a descendent may have a very different rate of evolution than its ancestor. In the second row, an autocorrelated clock, rates of evolution in the ancestor and descendent are expected to be more similar. As can be seen in the set of sample distributions, low values for the exponential rate parameter or the lognormal log variance parameter result in very wide distributions, implying that there can be a wide range of evolutionary rates across the tree. When the rate or log variance parameters are high, the rates are more constrained. The third row shows Dirichlet-distributed rates. This is a biologically agnostic clustering method for assigning branch rates. As can be seen in the distributions for this parameter, a high shape parameter implies a strong central tendency, and low values imply more variation in rates. Code to reproduce this figure is available online (DOI: 10.5281/zenodo.4035016).

5.4 Local Clocks

Random local clocks behave in some ways like strict clocks, and in some ways like relaxed clocks. A random local clock allows a subtree to have its own rate of evolution (Yoder & Yang, 2000). The branch subtending the subtree is the position of the shift between one clock rate and a new clock rate. Generally, the new clock rate applies to the whole subtree, without relaxation. The number of local clocks can vary between zero (one strict clock) to the number of branches on the tree (a fully relaxed clock). Both the number of clocks that describe the tree and the location of the shifts from one clock to another are sampled during the MCMC in implementations of this model (Drummond & Suchard, 2010). MCMC variants that allow the number of clocks to vary during inference, and therefore allow the total number of model parameters to vary between posterior samples, are called "reversible jump MCMC."

5.5 Other Models of Evolutionary Rate Variation

As described above, breaking up the branches of a tree into separate rate classes can be accomplished in many ways. Some have more straightforward biological interpretations, some have less. Another approach is to use a mixture model. Mixture models assume that there is substructure in a population of data. In this case, our population of data are branches that evolve under different rates. While the biological causes for those rates being different may not be the same, branches evolving under similar rates can be modeled together. Under a mixture model, the branches can be broken up into *n* categories. In the case that a strict clock is favored, *n* will be one category, or it can be many more under other circumstances.

Mixture models may be finite or infinite. In a finite mixture model, the number of different rates is specified a priori. In this case, while there is a defined number of categories, which branches belong to which categories is something that needs to be estimated. On the other hand, a mixture model may be infinite. In this case, the researcher does not specify a number of categories a priori: this is estimated during the phylogenetic inference (Heath, 2012). In these models, a Dirichlet Process Prior (DPP) is used to sample the number of categories, the average rate for each category, and which branches belong to each category. A DPP can be more concentrated (assuming fewer rate categories) or more diffuse (assuming more categories). Therefore, without assuming an explicit biological mechanism, they can be compatible with a number of biological scenarios.

6 Tree Models for Time-Calibrated Tree Inference

Tree models incorporate assumptions about the tree-generating processes and provide us with an expression for describing the probability of observing a given time-calibrated tree (see Fig. 2). This allows us to obtain a distribution containing the most likely trees, in terms of tree topology and branch durations, separate to any information we gain from the molecular or morphological character data. They also provide a framework for incorporating temporal evidence into our analyses – that is, we use the tree model to propose a plausible range of ages for the nodes in our phylogeny. In contrast to the substitution and clock models, only the tree model incorporates age information. This information is used to calibrate the substitution rate in combination with the substitution and clock model components.

Approaches to calibration can be placed into two useful categories: *node-dating* and *tip-dating*. These broadly reflect major differences in how age information is combined with or incorporated into the tree model. Briefly, node-dating assumes that our tree represents the relationships between living (extant) species only, and we constrain the ages of internal nodes using information from the geological record, without directly considering extinct or fossil samples as being part of the tree. In contrast, tip-dating directly considers fossil samples as part of the tree. In this section we provide an overview of popular tree models and describe how they are used in both node- or tip-dating scenarios.

The tree model is often referred to as the tree prior, and in combination with the calibration information, researchers often talk about the resulting prior distribution on node ages. Some of the inconsistency in terminology can be attributed to the history of different models used for phylogenetic dating and whether we consider age information used during inference (e.g., fossil sampling times) as data. Under the node-dating approach, fossil sampling times are used to constrain the age of a node. In this framework, they are not data because the generating process is not explicitly modeled. Instead, the fossil times are used to bound the age of a node. Alternatively, if we model the process of fossil recovery explicitly, it becomes clear that the fossil ages are actually data, in addition to the morphological characters. The terms process- and prior-based have also been used to distinguish between approaches that explicitly model the process that generated the temporal evidence used in our analysis and those that do not (Landis, 2017). Here, we use the term tree model to refer to all the models that underlie these different approaches. Tree models are a large and important family of models used in Bayesian divergence time inference. The tree model and/or the calibration information combined with the tree model can have a major impact on Bayesian estimates of node ages using both node- and tip-dating (e.g., Ho & Phillips, 2009; Matschiner et al., 2017; Matzke & Wright, 2016; O'Reilly, dos Reis, & Donoghue, 2015; Warnock et al., 2015).

6.1 Models of Speciation, Extinction, and Sampling

The most intuitive models are those that capture the processes we believe gave rise to our data and include parameters with tangible, biological meaning. An advantage of process-based tree models is that they can provide a better description of our data and also allow us to quantify other key parameters of interest, such as speciation (birth) and extinction (death) rates, in addition to the tree topology and divergence times. The most widely used tree models in macroevolution are birth-death process models, which refer to a huge family of models, at the heart of which are the speciation and extinction processes (together, known as diversification processes).

The simplest model, the pure-birth model, assumes speciation is constant over time, that we have no extinction, and that we sample a representative of every individual lineage (Yule, 1925). Under a pure-birth model with speciation rate λ, a single lineage splits in two with rate λ (with the expected time between events $= 1/\lambda$). Then, you have two lineages, each associated with rate λ, meaning you go from two to three lineages with rate 2λ. For any given number of lineages n, the rate of going from n to $n + 1$ will be $n\lambda$. The most straightforward extension incorporates extinction (Kendall, 1948). Similar to the birth process, a single lineage goes extinct with rate μ, meaning going from n to $n - 1$ lineages occurs with rate $n\mu$.

Restrictive assumptions, such as no extinction or constant rates of speciation, may be reasonable in small and recent clades, but are not likely to occur over long time intervals and for large groups. In reality, we hardly ever reach complete species sampling, especially in paleobiology. Some of the most important model developments in this area have therefore aimed to relax the assumption of complete sampling, both in the present and in the past. Sampling living species in the present and sampling either living or extinct species from the fossil record are typically treated as distinct processes. In particular, it is useful to think of extant species as being sampled in the present ($t = 0$) with a given probability ρ, which could be anywhere between 0 and 1, depending on the taxonomic scope of the study. In contrast, we tend to model fossil recovery as a continuous process, with an associated rate parameter ψ. Like the birth and death processes, a new fossil is recovered with rate $n\psi$.

Tree models capture the underlying processes (speciation, extinction, and sampling) that result in the *complete* tree, including sampled and nonsampled lineages. But to calculate the probability of observing the *reconstructed* tree (the tree representing the relationships between sampled individuals only), we need to account for the fact that we only sample some subset of lineages. For example, if we only sample living species, but assume both speciation and extinction have occurred, we need to use the expression for the probability of observing our tree, given we only sample species in the present and none in the past (Gernhard, 2008; Stadler, 2009; Thompson, 1975). Similarly, if we only sample a subset that does not include all living species, we need to use a model that incorporates incomplete extant species sampling (Yang & Rannala, 1997; Stadler, 2009). Fig. 5 shows examples of the complete versus the reconstructed tree for different birth-death process models.

The assumptions made by different tree models are important because they can result in very different distributions of plausible trees. Different combinations of the speciation, extinction, and sampling parameters give rise on average to different tree shapes, which determine the most probable waiting times between ancestor and descendent nodes in the reconstructed tree. For example, a reconstructed tree representing the relationships among a set of living individuals (i.e., the tree includes no extinct samples) is more likely to have shorter internal nonterminal branches and more evenly distributed speciation events if extinction is low relative to speciation. Conversely, the reconstructed tree is more likely to have longer internal branches and on average older node ages if extinction is high. More speciation events are missing from the reconstructed tree because extinct species are absent and there is a higher chance we have to go further back in time to find the speciation event linking any of our extant samples.

Note that we do not have to fix the speciation, extinction, and sampling parameters. Indeed, since different parameter combinations result in distinct distributions of trees and not all combinations are equally likely to result in the same tree shape, phylogenetic data allows us to estimate these parameters if they are explicitly part of the tree model. We typically use priors to constrain these parameters.

In the node-dating scenario, the tree represents the relationships between living samples and we typically use a tree model that includes extant species sampling only, excluding the process of fossil recovery. Temporal information from the fossil record is instead incorporated through the use of *node calibrations*. For one or more internal nodes in our phylogeny we may have information about the age of the speciation event based on fossil or other geological evidence. For example, for a given pair of lineages, the age of the first appearance of either one of these lineages represents a minimum (i.e., younger) bound for the age of the node separating them (Parham et al., 2012). We can represent the uncertainty in the age of this node using a probability distribution. This information is combined with the tree model to produce a distribution of trees that have branch lengths in units of absolute time. This approach is somewhat less biologically intuitive than an explicit model of diversification and fossil recovery, since it does not consider the process that gave rise to the data (i.e., the fossil sampling times). This leads to technical challenges in combining node calibrations with the tree model and in interpreting the resulting distribution on node ages (Heled & Drummond, 2012; Warnock et al., 2015). It also

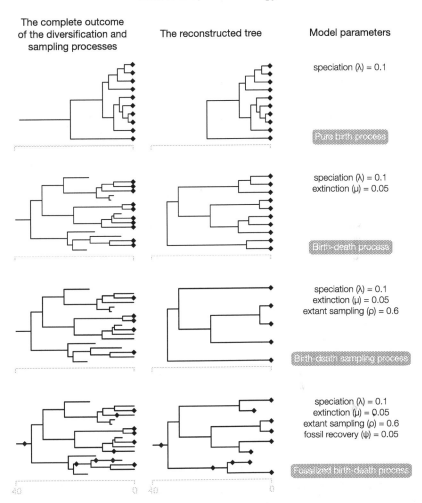

The complete outcome of the diversification and sampling processes   The reconstructed tree   Model parameters

speciation (λ) = 0.1

Pure birth process

speciation (λ) = 0.1
extinction (μ) = 0.05

Birth-death process

speciation (λ) = 0.1
extinction (μ) = 0.05
extant sampling (ρ) = 0.6

Birth-death sampling process

speciation (λ) = 0.1
extinction (μ) = 0.05
extant sampling (ρ) = 0.6
fossil recovery (ψ) = 0.05

Fossilized birth-death process

Figure 5 The complete versus reconstructed trees under birth-death process models. The assumptions of four different models are captured in each row. The first column shows an example outcome of the joint diversification and sampling processes (i.e., the complete tree), where diamonds represent extant or fossil samples. The second column shows the tree that contains sampled lineages only (i.e., the reconstructed tree). The third column shows the parameters and the name commonly applied to the model used to described the probability of observing the reconstructed tree shown in column 2, given we assume the generating processes shown in column 1. In all cases we assume constant speciation, extinction and fossil recovery, and uniform extant species sampling. Trees and fossils were simulated and plotted using the R packages TreeSim (Stadler, 2011) and FossilSim (Barido-Sottani et al., 2019). Code to reproduce this figure is available online (DOI: 10.5281/zenodo.4035016).

requires assigning a fossil age to a fixed node in the extant species tree, ignoring the potential for phylogenetic uncertainty in the placement of the fossil species.

In tip-dating we consider extinct samples explicitly as being part of the tree, and the temporal evidence used to constrain the age of the tree comes from the age of the extinct tips (Ronquist et al., 2012). To include fossil samples as part of the tree, we need to account for sampling through time, and ideally we want to use a tree model that incorporates the process of fossil recovery. The *fossilized birth-death* (FBD) *process* is an extension of the models described above that incorporates the fossil recovery process and provides an expression for the probability of observing a tree with samples recovered along internal branches (Stadler, 2010; Heath et al., 2014; Gavryushkina et al., 2014). Extinct samples can either occur on terminal branches (i.e., tips) or along branches leading to other sampled descendents, referred to as *sampled ancestors*.

When we consider fossil samples as being part of the tree-generating process, it becomes important to consider what each sample in our tree actually represents (Hopkins et al., 2018). In the fossil record, a species will be represented by one or more fossil occurrences. An occurrence could represent a single specimen or multiple specimens from the same locality. Further, the age of each occurrence will be associated with an age range, reflecting imprecision in dating techniques, which can be referred to as the *stratigraphic age* of an occurrence. This uncertainty can be accounted for by placing a prior distribution on the age of the fossil, instead of treating the age as a known variable (Drummond and Stadler, 2016; Barido-Sottani, Aguirre-Fernández et al., 2019). However, this is distinct from the observed duration of a species over geological time, beginning with the first (oldest) appearance of the species in the fossil record and terminating with the last (youngest) appearance, known as the *stratigraphic range* of a species. The FBD *range* process is more appropriate for incorporating information about species through time (Stadler et al., 2018). Birth-death process models have been extended in many ways, and of particular relevance to paleobiology, there are models that relax the assumption of uniform diversification or species sampling (Höhn, Stadler, et al., 2011; Stadler et al., 2013; Gavryushkina et al., 2014; Zhang et al., 2016; Kühnert et al., 2016; Barido-Sottani, Vaughan, & Stadler, 2020).

6.2 The Uniform Tree Model

Uniform tree models make the assumption that for a given set of taxa all possible trees are equally likely, and are available for both unconstrained and constrained (time-calibrated) tree inference (Huelsenbeck & Ronquist, 2001; Ronquist et al., 2012). For timetrees this model is used for tip-dating rather than node-dating (Ronquist et al., 2012). Fossil species are treated as extinct tips and sampled as part of the tree. Age information is incorporated through the fossil ages and an upper bound is applied to constrain the maximum age of the root. Internal node ages are drawn from a uniform distribution, satisfying the age constraints imposed by the root and tip ages. An advantage of this model is that it makes fewer explicit assumptions about both the diversification and the fossil and extant species sampling processes. In this sense, the uniform tree model is more straightforward, but has the disadvantage that it cannot be used to co-estimate diversification or sampling parameters.

In theory, given we have sufficient character data, the morphological data in combination with the terminal fossil ages should be informative about the substitution rate, and we should be able to recover the correct branch lengths, irrespective of the root constraint (Klopfstein et al., 2019; Ronquist et al., 2012). In reality, morphological datasets tend to be very small and this can result in the root constraint having a large impact on the results (Matzke & Wright, 2016). If the character data are not sufficiently informative about the substitution rate, we tend to observe that the older the root constraint, the older the node ages we recover, reflecting the uncertainty associated with the rate parameter. Although uniform tree models are sometimes referred to as uninformative tree priors, this is somewhat misleading if we consider the influence of the root constraint and the potential impact of ignoring sampled ancestors (Gavryushkina et al., 2014).

6.3 Coalescent Tree Models

Another large family of tree models used to describe the generation of time-trees are coalescent models. These are typically used to model the evolution of genes within a population, though they are also used in phylogenetic and phylogenomic estimation (Liu, 2008; Mirarab et al., 2014; Song et al., 2012). In this context, the tree typically represents a succession of nonoverlapping generations and each branching point represents a *coalescence event*, which is the point at which two genes in a population last shared a common ancestor (Kingman, 1982). In contrast to birth-death models, which are forward-time processes, coalescent models are backward-time processes. Time to coalescence will be a function of population size over time – the larger the population, the more likely you will have to go further back to recover the ancestor of two individuals. Similarly to birth-death models, coalescent models have also undergone an enormous amount of development, and provide flexible options for describing population growth (Beerli & Felsenstein, 2001; Drummond et al., 2005; Mashayekhi & Beerli, 2019).

Although coalescent models can incorporate extinct tips, we do not tend to use these directly to describe the evolution of species, but they can be important in estimating species trees and divergence times from genetic data. Trees based on individual genes can be quite different from the true underlying species history. This occurs when coalescence events between individuals belonging to populations of different species are older than the speciation event. This scenario is known as *incomplete lineage sorting* and can lead to a mismatch between gene and species trees. Following speciation, it takes time for genes to become sorted across distinct species populations, such that gene trees eventually reflect the species tree (Maddison & Knowles, 2006). This interval of time depends on several factors, including population size, and can be extremely long (e.g., populations of humans and chimpanzees still share genetic polymorphisms). However, the mismatch between gene and species trees can actually persist forever if genes do not become sorted before subsequent speciation events (Xu & Yang, 2016). Mismatch is most likely to occur when the branches separating speciation events are very short, irrespective of the time since speciation (i.e., whether the events are geologically recent or not). This scenario creates a huge challenge when inferring the species tree. Discerning the relationships between the major lineages of birds is a good example of this issue – these events happened almost 66 myr, but the internal branches in this portion of the tree are extremely short. As a consequence, different gene trees produce conflicting topologies (Jarvis et al., 2014). In the face of considerable conflict, identifying a consensus is not straightforward. One solution is to explicitly model the evolution of genes, in combination with the speciation process, under the *multi-species coalescent model* (Heled & Drummond, 2010). In this framework, we can apply a separate coalescent model to each gene in our dataset, and we model the speciation process using a birth-death model. We effectively assume that the gene trees are embedded within the species tree. We can use the FBD model for the species tree, meaning we can also incorporate extinct species, with or without molecular and/or morphological data (Ogilvie et al., 2018). If we also have morphological characters and assume that morphology follows the species tree history, rather than being described by a coalescent model, we can use the species tree model for the morphology. This is a good example of the hierarchical and extendable nature of phylogenetic tree models, but also showcases a level of complexity that will not always be necessary to recover the correct tree.

6.4 Biogeographic Dating

Temporal evidence for the age of a node can also come from the geological events linked to speciation (De Baets, Antonelli, & Donoghue, 2016; Ho et al., 2015). For example, the current biogeographic distribution of living taxa may indicate that species divergence is tied to specific tectonic events that likely resulted in genetic isolation, such as island formation or the breakup of continents that previously existed in Earth's history. This approach is especially useful for taxonomic groups with a sparse or nonexistent fossil record. Age information can be incorporated using a node dating approach, where the timing of biogeographic events is used to inform the calibration distributions, and the tree-generating process can be described using a birth-death model. One challenge to this approach is establishing a definitive causal link between tectonic and speciation processes, especially if events happened a long time ago (e.g., the breakup of Gondwana) (De Baets et al., 2016).

More recently, process-based models have been introduced for biogeographic dating and tree inference (Landis, 2017; Landis, Freyman, & Baldwin, 2019). This approach is conceptually similar to birth-death models that incorporate the fossil recovery process in that they explicitly incorporate a model of the evolution of biogeography. In this setup, we have information about the distribution of living species at the tips of our tree, and a model of tectonic history that incorporates age information. Species are allowed to disperse between areas with a given rate, which can depend on the current state of the tectonic configuration. For example, a species cannot disperse to an island before the island exists. Similarly, the potential for dispersal between two continents will depend on their connectivity. Thus, the probability of the tree and divergence times is linked to the biogeographic model. An advantage of this approach is that we do not need to make fixed assumptions about the link between biogeographic scenarios and speciation. Instead, we can use this approach to test among biogeographic hypotheses – not all histories will be equally likely to have produced the current distribution of living species. So far this approach has been used to date trees of extant species only; however, future extensions could potentially account for the biogeography of extinct and fossil samples. In principle, we could even combine models of biogeographic processes with models of diversification and fossil recovery.

7 Expanding the Potential of the Tripartite Model Within the Bayesian Framework

Bayesian priors incorporate our preexisting knowledge about parameter values. We tend to think about the role of priors as being restricted to constraining the range of possible values a given parameter can take, for example, the clock rate or speciation rate, or to express which values are most probable based on what we already know. However, we can use priors to manipulate the parameter space in much more sophisticated ways than we are currently used to doing. We can expand the range of assumptions we are able to make about the underlying biological processes and take advantage of Bayesian approaches to model testing. The development of more flexible Bayesian software, such as RevBayes (Höhna et al., 2014; Höhna et al., 2016) and BEAST2 (Bouckaert et al., 2019), alongside resources for understanding the underlying models (Barido-Sottani, Bošková et al., 2018), make complex inference much more accessible to everyday users. Here, we provide examples, ranging from simple to complex.

A tripartite model enables nearly endless combinations of substitution, clock, and tree models to be assembled into a complete model. For example, in molecular genetics, partitioning (defining subsets of the data) by gene and applying an appropriate model of sequence evolution to each gene, is strongly supported as being important to inferring a correct phylogeny (Brandley, Schmitz, & Reeder, 2005). Likewise, different models of morphological evolution can be used interchangeably. For example, the assumption made by the Mk model that a character is equally likely to change state as to reverse that state change may seem unrealistic. In this case, a model that allows asymmetrical rates of change (Ronquist, Huelsenbeck, & Britton, 2004) could be substituted. This alternative model puts a prior on character state frequencies, allowing them to be unequal, which increases the probability of certain types of character change. For example, many changes are likely to be observed from a common character state to other states. Character change asymmetry has recently been shown to affect divergence time estimation based on discrete character data (Klopfstein et al., 2019), as well as tree inference (Wright et al., 2016). If a researcher believes this to be the correct model for their data, it can be substituted for a traditional Mk model, without necessarily needing to alter the clock or tree models.

Depending on our parameters of interest, we can change the way our models are parameterized. For example, we may be more interested in diversification (d) and turnover (r) than speciation (λ) and extinction (μ). If we use a birth-death process tree model, we cannot eliminate speciation and extinction from the calculation, but we can reparameterize our analyses, such that we can place priors directly on the diversification and turnover parameters and sample these during MCMC. We can recover the speciation and extinction rates via transformation (Heath et al., 2014). The relationship between the parameters can be expressed simply as,

$$d = \lambda - \mu, \quad r = \frac{\mu}{\lambda} \quad \text{and}$$

$$\lambda = \frac{d}{(1 - r)}, \quad \mu = \frac{rd}{(1 - r)}.$$

Although in principle we can recover diversification and turnover from estimates of speciation and extinction without reparameterizing the model, this would give us less control over our parameters of interest. While this is a relatively straightforward example, this illustrates how parameters that are not explicitly part of the model can still be used to constrain the underlying model in our analysis. For instance, if other biological or environmental variables can be linked to model parameters via transformation, we have the potential to take advantage of this additional data.

We can also manipulate the relationship between independent parameters within a model through the use of priors. For example, we can link different parameters of the tree model in different ways. The FBD skyline model can incorporate variation in the speciation, extinction, and fossil recovery rates over time (Gavryushkina et al., 2014; Zhang et al., 2016). By default, model parameters are treated as independent. However, our prior assumption may be that parameters in adjacent time intervals, such as diversification rates, are more likely to be similar. To incorporate this expectation, the rate of diversification in a time interval could be parameterized according to the rate of diversification in the previous time interval, much like the relationship between descendent branches under the autocorrelated relaxed clock model. In effect, this allows for distinct time intervals to have semi-independent model parameters. Alternatively or in addition, if we have reason to believe that different model parameters are linked, we can also manipulate the priors to specify this expectation. For instance, we may have reason to believe that rates of diversification are linked to the rate of fossil sampling (Holland, 1995; Peters, 2005).

However, we may very well believe that parameters in *different* subcomponents of the model are linked. If we believed the rate of speciation to be related to the rate of character change, this could be achieved by using a prior that specifies a distribution for one parameter, centered on the other. For example, if we thought that periods of high speciation would correspond to periods with lots of character change, we could create an FBD skyline model in which the per-interval prior on speciation rate is linked to the average substitution rate during that interval.

Within a Bayesian framework, we can propose any model we like, and use modeling tests to compare competing models, in which parameters are either linked or not. Bayesian methods have a suite of well-developed statistical approaches for evaluating the fit of both the model and the priors to the data. For instance, Bayes Factors (Xie et al., 2011) are metrics that describe the support for one model, and all its associated priors, over another model. This approach weighs the posterior evidence of two models against one another. It is worth noting, however, that the Bayes Factor can only provide evidence in favor of one model. It cannot tell a researcher whether the model is adequate; that is, that it captures important facets of the process of evolution. Other methods, such as posterior predictive model assessment, can be used to assess model adequacy (Brown & ElDabaje, 2009; Brown, 2014; Duchêne et al., 2015; Höhna et al., 2017). With these methods, it is important to consider what the data are. Node calibration methods, for example do not truly incorporate fossils as data. Instead, the fossils are used as priors to bound the age of nodes. In this case, their placement is part of the model, and methods have been proposed to evaluate these priors with Bayes Factors (Andújar et al., 2014). In the case of an FBD tree model, they are data, thus Bayes Factor model fitting cannot be used to evaluate the placement of fossils.

Many of the more complex model and prior options we describe here have yet to be explored using paleobiological data, despite their increasing feasibility. To extend the tripartite model, we must understand how it works under a variety of empirical conditions. Much of what we know about both divergence time estimation and phylogenetic analysis comes from simulation studies and mathematical modeling. While both of these are useful tools, it can be difficult to understand how the behavior of any particular method will stand up to empirical conditions. Limited data sizes, biased missing data, and violations of model assumptions can all lead to unpredictable analytical behavior. Therefore, it is critical for empiricists and theoreticians to collaborate in order to understand the challenges faced by researchers at the forefront of collecting data, and improve our methods to meet them.

Here, we focused on the scenario in which we have both phylogenetic character data and temporal evidence of speciation, and where the goal is to estimate divergence times or to co-estimate divergence times and topology. The inclusion of molecular or morphological characters requires both the substitution and clock models. We cannot infer the topology without phylogenetic characters; however, in principle, the tree model could still be used to infer the divergence times for a tree topology obtained using other evidence. For example, phylogenies based on taxonomic classification have been shown to be valuable in phylogenetic comparative analyses (Soul & Friedman, 2015). In this context, the tree may not be fully resolved, but the timing of key divergence events can still be estimated under the tree model, taking into account the uncertainty at unresolved nodes. Similarly, since fossil sampling times are informative about the speciation, extinction, and fossil recovery rates, the FBD model (Warnock, Heath, & Stadler, 2020) and related birth-death models (Silvestro et al., Liow, Antonelli, & Salamin, 2014) can be used to recover these parameters, even without any knowledge of the underlying phylogeny.

8 Conclusions

Bayesian divergence time estimation is commonly performed in a tripartite framework. One model describes the process the researcher believes generated the character data. Another model describes the manner in which the researcher believes rates of evolution are distributed across the tree. The final model describes the extinction, speciation, and sampling events that potentially led to the observed tree. Each of these components has its own parameters, which are believed to describe the process that generated the data. Each component model's parameters can have priors too, which describe the distribution of values we expect a parameter to take.

This framework enables nearly endless combinations of assumptions that a researcher can make about their data. The goal of this review has been to explain some common assumptions, and what they mean. It is by no means exhaustive. There are more assumptions that could be made and modeled by researchers. This tripartite framework can be improved by a close collaboration between geologists, organismal experts, and phylogenetic methods specialists. We hope that, as a result of our explaining some of these common assumptions, researchers will feel empowered to look at their own data and see where methods can be improved, and to seek collaborations to create a new generation of process-driven methods. The challenge for both empirical researchers and method developers will be to identify important model violations, and to gauge the level of complexity required to obtain reliable and meaningful results.

---

1. THE LIKELIHOOD, THE PRIOR, AND THE POSTERIOR

It can be confusing at first to understand what the model likelihood, the prior, and the posterior truly mean. In plain language, the model likelihood is the probability of the data given a model. Without a model, there can be no calculation of the model likelihood.

Priors can be set on parameters in the model, specifying distributions from which the value is thought to be drawn. These distributions are often based on the researcher's intuition, and on information from prior studies.

The posterior distribution is a set of plausible solutions given the model likelihood and the prior. During Bayesian estimation, different values will be sampled for model parameters. Their probability will be evaluated according to the likelihood and the prior. Therefore, the posterior is proportional to the likelihood and the prior. A good solution will often appear in the posterior sample many times.

In phylogenetics, we often refer to our models as continuous-time Markov chains. "Continuous-time" refers to models allowing change between character states to occur instantaneously at any point in an evolutionary history. Changes in the character state are not confined to the node; instead, branch lengths on a phylogeny are proportional to the number of expected changes per character along that particular branch. In this context, "Markov chain" refers to the fact that (existing) phylogenetic models are memory-less. This means the joint probability – taking into account all the parameters for the model of morphological substitution, the model of molecular substitution, and the tree and clock models – depends only on the current state and not on the history of the process. In practice, this is the computer model that we use to estimate the posterior (Höhna et al., 2016).

HIERARCHICAL MODELS

The tripartite approach to divergence time estimation is what is termed a *hierarchical model*. Hierarchical models are models in which variation may be described by different submodels. In the case of divergence time estimation, the character data (molecular and/or morphological) are described by one model, such as the Mk model. The distribution of evolutionary rates across branches is described by the clock model. Finally, the distribution of speciation, extinction, and fossil sampling is described by the tree model. Together, these three components are used to estimate a tree, branch lengths in units of time, and other relevant model parameters.

This term may be confusing, as model components may have a hierarchy of priors. For example, if we placed a lognormal distribution with shape parameter 10 on the mean clock rate, this is a prior. If instead, we placed an exponential prior on the shape parameter of the lognormal distribution, that exponential prior is called *hyperprior*. This, while a hierarchy of priors, is not a hierarchical model in the same way that the complete tripartite model for divergence time estimation is hierarchical.

See Heath (2012) for a nice example of the hyperprior approach to modeling uncertainty in the parameters associated with fossil calibration densities.

MAXIMUM LIKELIHOOD AND BAYESIAN ESTIMATION

As discussed in the Box "The Likelihood, the Prior, and the Posterior," the probability of the data is calculated given a model. In maximum likelihood estimation, models are proposed, and the likelihood of the data is calculated given each of those models. The model that gives the highest likelihood is considered to be "the best." This is generally a point estimate returning one tree, one set of branch lengths, and one set of other model parameters. See Holder and Lewis (2003) and Yang and Rannala (2012) for more information on the details and history of these different approaches.

Inference of undated trees from molecular and/or morphological data can be accomplished in many pieces of maximum-likelihood software, such as PAUP (Swofford, 2003), RAxML (Stamatakis, 2014), IQTREE (Nguyen et al., 2014), and GARLI (Zwickl, 2006). Estimation of dated trees incorporating molecular and/or morphological data has mostly been accomplished in a Bayesian context, using software such as MrBayes (Huelsenbeck et al., 2002; Ronquist & Huelsenbeck, 2003), BEAST (Suchard et al., 2018), BEAST2 (Bouckaert et al., 2019), MCMCTree (Yang, 2007), and RevBayes (Höhna et al., 2014; Höhna et al., 2016). While there is no reason models such as the FBD cannot be estimated using maximum likelihood, in practice, it is not straightforward to incorporate the uncertainty associated with parameters within a maximum-likelihood framework.

References

Abadi, S., Azouri, D., Pupko, T., & Mayrose, I. (2019). Model selection may not be a mandatory step for phylogeny reconstruction. *Nature Communications, 10*(1), 934.

Álvarez-Carretero, S., Goswami, A., Yang, Z., & dos Reis, M. (2019). Bayesian estimation of species divergence times using correlated quantitative characters. *Systematic Biology,* syz015. doi: https://doi.org/10.1093/sysbio/syz015

Andújar, C., Soria-Carrasco, V., Serrano, J., & Gómez-Zurita, J. (2014). Congruence test of molecular clock calibration hypotheses based on Bayes factor comparisons. *Methods in Ecology and Evolution, 5*(3), 226–242. doi: https://besjournals.onlinelibrary.wiley.com/doi/abs/10.1111/2041-210X.12151

Aris-Brosou, S., & Yang, Z. (2002). Effects of models of rate evolution on estimation of divergence dates with special reference to the metazoan 18S ribosomal RNA phylogeny. *Systematic Biology, 51*(5), 703–714.

Aris-Brosou, S., & Yang, Z. (2003). Bayesian models of episodic evolution support a late Precambrian explosive diversification of the Metazoa. *Molecular Biology and Evolution, 20*(12), 1947–1954.

Bapst, D. W., Schreiber, H. A., & Carlson, S. J. (2017). Combined analysis of extant Rhynchonellida (Brachiopoda) using morphological and molecular data. *Systematic Biology, 67*(1), 32–48. doi: `https://doi.org/10.1093/sysbio/syx049`

Barido-Sottani, J., Aguirre-Fernández, G., Hopkins, M. J., Stadler, T., & Warnock, R. C. M. (2019). Ignoring stratigraphic age uncertainty leads to erroneous estimates of species divergence times under the fossilized birth–death process. *Proceedings of the Royal Society B, 286*(1902), 20190685.

Barido-Sottani, J., Bošková, V., du Plessis, L., et al. (2018). Taming the beast? A community teaching material resource for BEAST 2. *Systematic Biology, 67*(1), 170–174.

Barido-Sottani, J., Pett, W., O'Reilly, J. E., & Warnock, R. C. M. (2019). FossilSim An R package for simulating fossil occurrence data under mechanistic models of preservation and recovery. *Methods in Ecology and Evolution, 10*(6), 835–840.

Barido-Sottani, J., Vaughan, T. G., & Stadler, T. (2020). A multi-state birth-death model for Bayesian inference of lineage-specific birth and death rates. *Systematic Biology, 69*(5), 973–986. doi: https://doi.org/10.1093/sysbio/syaa01

Beaulieu, J. M., Jhwueng, D.-C., Boettiger, C., & O'Meara, B. C. (2012). Modeling stabilizing selection: Expanding the Ornstein–Uhlenbeck model of adaptive evolution. *Evolution: International Journal of Organic Evolution, 66*(8), 2369–2383.

Beerli, P., & Felsenstein, J. (2001). Maximum likelihood estimation of a migration matrix and effective population sizes in n subpopulations by using a coalescent approach. *Proceedings of the National Academy of Sciences, 98*(8), 4563–4568. doi: https://doi.org/10.1073/pnas.081068098

Bouckaert, R. R., Vaughan, T. G., Barido-Sottani, J., et al. (2019). BEAST 2.5: An advanced software platform for Bayesian evolutionary analysis. *PLoS Computational Biology, 15*(4), 1–28. doi: `https://doi.org/10.1371/journal.pcbi.1006650`

Brandley, M. C., Leaché, A. D., Warren, D. L., & McGuire, J. A. (2006). Are unequal clade priors problematic for Bayesian phylogenetics? *Systematic Biology, 55*(1), 138–146.

Brandley, M. C., Schmitz, A., & Reeder, T. W. (2005). Partitioned Bayesian analyses, partition choice, and the phylogenetic relationships of scincid lizards. *Systematic Biology, 54*(3), 373–390.

Bromham, L., Hua, X., Lanfear, R., & Cowman, P. F. (2015). Exploring the relationships between mutation rates, life history, genome size, environment, and species richness in flowering plants. *The American Naturalist, 185*(4), 507–524.

Bromham, L., Rambaut, A., & Harvey, P. H. (1996). Determinants of rate variation in mammalian DNA sequence evolution. *Journal of Molecular Evolution, 43*(6), 610–621.

Brown, J., & ElDabaje, R. (2009). PuMA: Bayesian analysis of partitioned (and unpartitioned) model adequacy. *Bioinformatics, 25*(4), 537–538.

Brown, J. M. (2014). Predictive approaches to assessing the fit of evolutionary models. *Systematic Biology, 63*(3), 289–292.

Butler, M. A., & King, A. A. (2004). Phylogenetic comparative analysis: A modeling approach for adaptive evolution. *The American Naturalist, 164*(6), 683–695.

De Baets, K., Antonelli, A., & Donoghue, P. C. (2016). Tectonic blocks and molecular clocks. *Philosophical Transactions of the Royal Society B: Biological Sciences, 371*(1699), 20160098.

de Queiroz, K. (1985). The ontogenetic method for determining character polarity and its relevance to phylogenetic systematics. *Systematic Zoology, 34*(3), 280–299.

dos Reis, M., Donoghue, P. C., & Yang, Z. (2016). Bayesian molecular clock dating of species divergences in the genomics era. *Nature Reviews Genetics, 17*(2), 71.

dos Reis, M., & Yang, Z. (2013). The unbearable uncertainty of Bayesian divergence time estimation. *Journal of Systematics and Evolution, 51*(1), 30–43.

Drummond, A. J., Ho, S. Y., Phillips, M., & Rambaut, A. (2006). Relaxed phylogenetics and dating with confidence. *PLoS Biology, 4*(5), e88.

Drummond, A. J., & Rambaut, A. (2007). BEAST: Bayesian evolutionary analysis sampling trees. *BMC Evolutionary Biology, 7*, 214.

Drummond, A. J., & Suchard, M. (2010). Bayesian random local clocks, or one rate to rule them all. *BMC Biology, 8*(1), 114.

Drummond, A. J., Rambaut, A., Shapiro, B., & Pybus, O. G. (2005). Bayesian coalescent inference of past population dynamics from molecular sequences. *Molecular Biology and Evolution, 22*(5), 1185–1192. doi: `https://doi.org/10.1093/molbev/msi103`

Drummond, A. J., & Stadler, T. (2016). Bayesian phylogenetic estimation of fossil ages. *Philosophical Transactions of the Royal Society B: Biological Sciences, 371*(1699), 20150129.

Duchêne, D. A., Duchêne, S., Holmes, E. C., & Ho, S. Y. (2015). Evaluating the adequacy of molecular clock models using posterior predictive simulations. *Molecular Biology and Evolution, 32*(11), 2986–2995.

du Plessis, L., & Stadler, T. (2015). Getting to the root of epidemic spread with phylodynamic analysis of genomic data. *Trends in Microbiology, 23*(7), 383–386.

Felsenstein, J. (1973). Maximum-likelihood estimation of evolutionary trees from continuous characters. *American Journal of Human Genetics, 25*(5), 471–92.

Felsenstein, J. (1981). Evolutionary trees from DNA sequences: A maximum likelihood approach. *Journal of Molecular Evolution, 17*(6), 368–376.

Felsenstein, J. (1985). Phylogenies and the comparative method. *The American Naturalist*, 1–15.

Gaut, B. S., Muse, S. V., Clark, W. D., & Clegg, M. T. (1992). Relative rates of nucleotide substitution at the *rbcL* locus of monocotyledonous plants. *Journal of Molecular Evolution, 35*(4), 292–303.

Gavryushkina, A., Heath, T. A., Ksepka, D. T., et al. (2017). Bayesian total-evidence dating reveals the recent crown radiation of penguins. *Systematic Biology, 66*(1), 57–73.

Gavryushkina, A., Welch, D., Stadler, T., & Drummond, A. J. (2014). Bayesian inference of sampled ancestor trees for epidemiology and fossil calibration. *PLoS Computational Biology, 10*(12), e1003919.

Gernhard, T. (2008). The conditioned reconstructed process. *Journal of Theoretical Biology, 253*(4), 769–778.

Gingerich, P. D. (1993). Quantification and comparison of evolutionary rates. *American Journal of Science, 293*(A), 453.

Goloboff, P. A., Mattoni, C. I., & Quinteros, A. S. (2006). Continuous characters analyzed as such. *Cladistics, 22*(6), 589–601.

Hansen, T. F. (1997). Stabilizing selection and the comparative analysis of adaptation. *Evolution, 51*(5), 1341–1351.

Harrison, L. B., & Larsson, H. C. (2015). Among-character rate variation distributions in phylogenetic analysis of discrete morphological characters. *Systematic Biology, 64*(2), 307–324.

Hasegawa, M., Kishino, H., & Yano, T. (1985). Dating of the human-ape splitting by a molecular clock of mitochondrial DNA. *Journal of Molecular Evolution, 22*(2), 160–174.

Heath, T. A. (2012). A hierarchical Bayesian model for calibrating estimates of species divergence times. *Systematic Biology, 61*(5), 793–809.

Heath, T. A., Huelsenbeck, J. P., & Stadler, T. (2014). The fossilized birth-death process for coherent calibration of divergence-time estimates. *Proceedings of the National Academy of Sciences, 111*(29), E2957–E2966.

Heled, J., & Bouckaert, R. R. (2013). Looking for trees in the forest: Summary tree from posterior samples. *BMC Evolutionary Biology, 13*(1), 221.

Heled, J., & Drummond, A. J. (2010). Bayesian inference of species trees from multilocus data. *Molecular Biology and Evolution, 27*(3), 570.

Heled, J., & Drummond, A. J. (2012). Calibrated tree priors for relaxed phylogenetics and divergence time estimation. *Systematic Biology, 61*(1), 138–149.

Ho, S. Y., & Phillips, M. J. (2009). Accounting for calibration uncertainty in phylogenetic estimation of evolutionary divergence times. *Systematic Biology, 58*(3), 367–380.

Ho, S. Y., Tong, K. J., Foster, C. S., et al. (2015). Biogeographic calibrations for the molecular clock. *Biology Letters, 11*(9), 20150194.

Höhna, S., Coghill, L. M., Mount, G. G., Thomson, R. C., & Brown, J. M. (2017). P3: Phylogenetic posterior prediction in RevBayes. *Molecular Biology and Evolution, 35*(4), 1028–1034. doi: https://doi.org/10.1093/molbev/msx286

Höhna, S., Heath, T. A., Boussau, B., et al. (2014). Probabilistic graphical model representation in phylogenetics. *Systematic Biology, 63*(5), 753–771. doi: https://doi.org/10.1093/sysbio/syu039

Höhna, S., Landis, M. J., Heath, T. A., et al. (2016). RevBayes: Bayesian phylogenetic inference using graphical models and an interactive model-specification language. *Systematic Biology, 65*(4), 726–736. doi: https://doi.org/10.1093/sysbio/syw021

Höhn, S., Stadler, T., Ronquist, F., & Britton, T. (2011). Inferring speciation and extinction rates under different species sampling schemes. *Molecular Biology and Evolution, 28*(9), 2577–2589.

Holder, M., & Lewis, P. O. (2003). Phylogeny estimation: Traditional and Bayesian approaches. *Nature Reviews Genetics, 4*(4), 275.

Holland, S. M. (1995). The stratigraphic distribution of fossils. *Paleobiology, 21*(1), 92–109. doi: https://doi.org/10.1017/S0094837300013099

Hopkins, M. J., Bapst, D. W., Simpson, C., & Warnock, R. C. M. (2018). The inseparability of sampling and time and its influence on attempts to unify the molecular and fossil records. *Paleobiology, 44*(4), 561–574.

Huelsenbeck, J. P., Larget, B., Miller, R., & Ronquist, F. (2002). Potential applications and pitfalls of Bayesian inference of phylogeny. *Systematic Biology, 51*(5), 673–688.

Huelsenbeck, J. P., & Ronquist, F. (2001). MRBAYES: Bayesian inference of phylogenetic trees. *Bioinformatics, 17*(8), 754–755.

Jarvis, E. D., Mirarab, S., Aberer, A. J., et al. (2014). Whole-genome analyses resolve early branches in the tree of life of modern birds. *Science, 346*(6215), 1320–1331.

Jukes, T., & Cantor, C. (1969). Evolution of protein molecules. *Mammalian Protein Metabolism, 3*, 21–132.

Kendall, D. G. (1948). On the generalized "birth-and-death" process. *The Annals of Mathematical Statistics, 19*(1), 1–15.

Kimura, M. (1980). A simple method for estimating evolutionary rates of base substitutions through comparative studies of nucleotide sequences. *Journal of Molecular Evolution, 16*(2), 111–120.

King, B., Qiao, T., Lee, M. S. Y., Zhu, M., & Long, J. A. (2016). Bayesian morphological clock methods resurrect placoderm monophyly and reveal rapid early evolution in jawed vertebrates. *Systematic Biology, 66*(4), 499–516. doi: https://doi.org/10.1093/sysbio/syw107

Kingman, J. F. C. (1982). On the genealogy of large populations. *Journal of Applied Probability, 19*, 27–43.

Kishino, H., Thorne, J. L., & Bruno, W. J. (2001). Performance of a divergence time estimation method under a probabilistic model of rate evolution. *Molecular Biology and Evolution, 18*(3), 352–361.

Klopfstein, S., Ryser, R., Corio, M., & Spasejovic, T. (2019). Mismatch of the morphology model is mostly unproblematic in total-evidence dating: Insights from an extensive simulation study. *bioRxiv*, 679084.

Kühnert, D., Stadler, T., Vaughan, T. G., & Drummond, A. J. (2016). Phylodynamics with migration: A computational framework to quantify population structure from genomic data. *Molecular Biology and Evolution, 33*(8), 2102–2116.

Landis, M. J. (2017). Biogeographic dating of speciation times using paleogeographically informed processes. *Systematic Biology, 66*(2), 128–144. doi: https://doi.org/10.1093/sysbio/syw040

Landis, M. J., Freyman, W. A., & Baldwin, B. G. (2019). Retracing the Hawaiian silversword radiation despite phylogenetic, biogeographic, and paleogeographic uncertainty. *Evolution, 72*(11), 2343–2359.

Landis, M. J., Schraiber, J. G., & Liang, M. (2013). Phylogenetic analysis using Lévy processes: Finding jumps in the evolution of continuous traits. *Systematic Biology, 62*(2), 193–204.

Leaché, A. D., Banbury, B. L., Felsenstein, J., de Oca, A. n. M., & Stamatakis, A. (2015). Short tree, long tree, right tree, wrong tree: New acquisition bias corrections for inferring SNP phylogenies. *Systematic Biology, 64*(6), 1032–1047.

Lee, M.S.Y., Cau, A., Naish, D., & Dyke, G.J. (2014). Morphological clocks in paleontology, and a mid-Cretaceous origin of crown Aves. *Systematic Biology, 63*(3), 442–449. doi: https://doi.org/10.1093/sysbio/syt110

Lewis, P. O. (2001). A likelihood approach to estimating phylogeny from discrete morphological character data. *Systematic Biology, 50*(6), 913–925.

Liu, L. (2008). BEST: Bayesian estimation of species trees under the coalescent model. *Bioinformatics, 24*(21), 2542–2543.

Maddison, W. P., & Knowles, L. L. (2006). Inferring phylogeny despite incomplete lineage sorting. *Systematic Biology, 55*(1), 21–30.

Mashayekhi, S., & Beerli, P. (2019). Fractional coalescent. *Proceedings of the National Academy of Sciences, 116*(13), 6244–6249. www.pnas.org/content/116/13/6244 doi: https://doi.org/10.1073/pnas.1810239116

Matschiner, M., Musilová, Z., Barth, J. M., Starostová, Z., Salzburger, W., Steel, M., & Bouckaert, R. R. (2017). Bayesian phylogenetic estimation of clade ages supports trans-atlantic dispersal of cichlid fishes. *Systematic Biology, 66*(1), 3–22.

Matzke, N. J., & Wright, A. (2016). Inferring node dates from tip dates in fossil Canidae: The importance of tree priors. *Biology Letters, 12*(8), 20160328.

Mirarab, S., Reaz, R., Bayzid, M. S., Zimmermann, T., Swenson, M. S., & Warnow, T. (2014). ASTRAL: Genome-scale coalescent-based species tree estimation. *Bioinformatics, 30*(17), i541–i548.

Nascimento, F. F., dos Reis, M., & Yang, Z. (2017). A biologist's guide to Bayesian phylogenetic analysis. *Nature Ecology & Evolution, 1*(10), 1446–1454.

Nguyen, L.-T., Schmidt, H. A., von Haeseler, A., & Minh, B. Q. (2014). IQ-TREE: A fast and effective stochastic algorithm for estimating maximum-likelihood phylogenies. *Molecular Biology and Evolution, 32*(1), 268–274. doi: https://doi.org/10.1093/molbev/msu300

Nylander, J. A., Ronquist, F., Huelsenbeck, J. P., & Nieves-Aldrey, J. (2004). Bayesian phylogenetic analysis of combined data. *Systematic Biology, 53*(1), 47–67.

Ogilvie, H. A., Vaughan, T. G., Matzke, N. J. et al. (2018). Inferring species trees using integrative models of species evolution. *bioRxiv*, 242875.

O'Reilly, J. E., & Donoghue, P. C. (2018). The efficacy of consensus tree methods for summarizing phylogenetic relationships from a posterior sample of trees estimated from morphological data. *Systematic Biology, 67*(2), 354–362.

O'Reilly, J. E., dos Reis, M., & Donoghue, P. C. (2015). Dating tips for divergence-time estimation. *Trends in Genetics, 31*(11), 637–650.

Parham, J. F., Donoghue, P. C., Bell, C. J., et al. (2012). Best practices for justifying fossil calibrations. *Systematic Biology, 67*(2), 346–359.

Parins-Fukuchi, C. (2017). Use of continuous traits can improve morphological phylogenetics. *Systematic Biology, 67*(2), 328–339.

Paterson, J. R., Edgecombe, G. D., & Lee, M. S. Y. (2019). Trilobite evolutionary rates constrain the duration of the Cambrian explosion. *Proceedings of the National Academy of Sciences, 116*(10), 4394–4399.

Peters, S. E. (2005, Aug.). Geologic constraints on the macroevolutionary history of marine animals. *Proceedings of the National Academy of Sciences, 102*(35), 12326–12331. doi: https://doi.org/10.1073/pnas.0502616102

Ronquist, F., & Huelsenbeck, J. P. (2003). MrBayes 3: Bayesian phylogenetic inference under mixed models. *Bioinformatics, 19*(12), 1572–1574.

Ronquist, F., Huelsenbeck, J. P., & Britton, T. (2004). Bayesian supertrees. *Phylogenetic Supertrees: Combining Information to Reveal the Tree of Life, 3*, 193–224.

Ronquist, F., Klopfstein, S., Vilhelmsen, L., et al. (2012). A total-evidence approach to dating with fossils, applied to the early radiation of the Hymenoptera. *Systematic Biology, 61*(6), 973–999.

Schrago, C., Mello, B., & Soares, A. (2013). Combining fossil and molecular data to date the diversification of New World primates. *Journal of Evolutionary Biology, 26*(11), 2438–2446.

Silvestro, D., Schnitzler, J., Liow, L. H., Antonelli, A., & Salamin, N. (2014). Bayesian estimation of speciation and extinction from incomplete fossil occurrence data. *Systematic Biology, 63*(3), 349–367, doi: https://doi.org/10.1093/sysbio/syu006.

Slater, G. J. (2015). Iterative adaptive radiations of fossil canids show no evidence for diversity-dependent trait evolution. *Proceedings of the National Academy of Sciences, 112*(16), 4897–4902. www.pnas.org/content/112/16/4897 doi: https://doi.org/10.1073/pnas.1403666111

Song, S., Liu, L., Edwards, S. V., & Wu, S. (2012). Resolving conflict in eutherian mammal phylogeny using phylogenomics and the multispecies coalescent model. *Proceedings of the National Academy of Sciences, 109*(37), 14942–14947.

Soul, L. C., & Friedman, M. (2015). Taxonomy and phylogeny can yield comparable results in comparative paleontological analyses. *Systematic Biology, 64*(4), 608–620.

Stadler, T. (2009). On incomplete sampling under birth-death models and connections to the sampling-based coalescent. *Journal of Theoretical Biology, 261*(1), 58–66.

Stadler, T. (2010). Sampling-through-time in birth-death trees. *Journal of Theoretical Biology, 267*(3), 396–404.

Stadler, T. (2011). Mammalian phylogeny reveals recent diversification rate shifts. *Proceedings of the National Academy of Sciences, 108*(15), 6187–6192.

Stadler, T., Gavryushkina, A., Warnock, R. C. M., Drummond, A. J., & Heath, T. A. (2018). The fossilized birth-death model for the analysis of stratigraphic range data under different speciation modes. *Journal of Theoretical Biology, 447*, 41–55.

Stadler, T., Kühnert, D., Bonhoeffer, S., & Drummond, A. J. (2013). Birth-death skyline plot reveals temporal changes of epidemic spread in HIV and hepatitis C virus (HCV). *Proceedings of the National Academy of Sciences, 110*(1), 228–233.

Stamatakis, A. (2014). RAxML version 8: A tool for phylogenetic analysis and post-analysis of large phylogenies. *Bioinformatics, 30*(9), 1312–1313. doi: https://doi.org/10.1093/bioinformatics/btu033

Suchard, M. A., Lemey, P., Baele, G., et al. (2018). Bayesian phylogenetic and phylodynamic data integration using BEAST 1.10. *Virus Evolution, 4*(1). doi: https://doi.org/10.1093/ve/vey016 doi: 10.1093/ve/vey016

Swofford, D. L. (2003). PAUP*. Phylogenetic analysis using parsimony (*and other methods). Version 4. Sinauer Associates, Sunderland, Massachusetts. doi: https://paup.phylosolutions.com/documentation/faq/

Tavaré, S. (1986). Some probabilistic and statistical problems in the analysis of DNA sequences. *Some Mathematical Questions in Biology: DNA Sequence Analysis, 17*, 57–86.

Thomas, J. A., Welch, J. J., Woolfit, M., & Bromham, L. (2006). There is no universal molecular clock for invertebrates, but rate variation does not scale with body size. *Proceedings of the National Academy of Sciences, 103*(19), 7366–7371.

Thompson, E. (1975). *Human evolutionary trees.* Cambridge: Cambridge University Press.

Thorne, J. L., Kishino, H., & Painter, I. S. (1998). Estimating the rate of evolution of the rate of molecular evolution. *Molecular Biology and Evolution, 15*, 1647–1657.

Thorne, J. L., & Kishino, H. (2002). Divergence time and evolutionary rate estimation with multilocus data. *Systematic Biology, 51*(5), 689–702.

Wagner, P. J. (2011). Modelling rate distributions using character compatibility: Implications for morphological evolution among fossil invertebrates. *Biology Letters, 8*(1), 143–146.

Warnock, R. C. M., Heath, T. A., & Stadler, T. (2020). Assessing the impact of incomplete species sampling on estimates of speciation and extinction rates. *Paleobiology, 46*(2), 137–157, doi: https://doi.org/10.1017/pab.2020.12

Warnock, R. C. M., Parham, J. F., Joyce, W. G., Lyson, T. R., & Donoghue, P. C. (2015). Calibration uncertainty in molecular dating analyses: There is no substitute for the prior evaluation of time priors. *Proceedings of the Royal Society B: Biological Sciences, 282*(1798), 20141013.

Warnock, R. C. M., Yang, Z., & Donoghue, P. C. (2017). Testing the molecular clock using mechanistic models of fossil preservation and molecular evolution. *Proceedings of the Royal Society B: Biological Sciences, 284*(1857), 20170227.

Watrous, L. E., & Wheeler, Q. D. (1981). The out-group comparison method of character analysis. *Systematic Biology, 30*(1), 1–11.

Wood, H. M., Matzke, N. J., Gillespie, R. G., & Griswold, C. E. (2013). Treating fossils as terminal taxa in divergence time estimation reveals ancient vicariance patterns in the palpimanoid spiders. *Systematic Biology, 62*(2), 264–284.

Wright, A. M. (2019). A Systematist's guide to estimating bayesian phytogenies from morphological data. *Insect Systematics and Diversity, 3*(3). doi: https://doi.org/10.1093/isd/ixz006

Wright, A. M., Lloyd, G. T., & Hillis, D. M. (2016). Modeling character change heterogeneity in phylogenetic analyses of morphology through the use of priors. *Systematic Biology, 65*(4), 602–611.

Wright, D. F. (2017). Phenotypic innovation and adaptive constraints in the evolutionary radiation of Palaeozoic crinoids. *Scientific Reports, 7*(1), 13745. doi: https://doi.org/10.1038/s41598-017-13979-9

Wright, D. F., & Toom, U. (2017). New crinoids from the Baltic region (Estonia): Fossil tip-dating phylogenetics constrains the origin and Ordovician–Silurian diversification of the Flexibilia (Echinoder-mata). *Palaeontology, 60*(6), 893–910. doi: `https://onlinelibrary.wiley.com/doi/abs/10.1111/pala.12324`

Xie, W., Lewis, P. O., Fan, Y., Kuo, L., & Chen, M. (2011). Improving marginal likelihood estimation for Bayesian phylogenetic model selection. *Systematic Biology, 60*(2), 150–160.

Xu, B., & Yang, Z. (2016). Challenges in species tree estimation under the multispecies coalescent model. *Genetics, 204*(4), 1353–1368.

Yang, Z. (1994). Maximum likelihood phylogenetic estimation from DNA sequences with variable rates over sites: Approximate methods. *Journal of Molecular Evolution, 39*(3), 306–314.

Yang, Z. (2007). PAML 4: Phylogenetic analysis by maximum like-lihood. *Molecular Biology and Evolution, 24*(8), 1586–1591. doi: `https://doi.org/10.1093/molbev/msm088`

Yang, Z., & Rannala, B. (1997). Bayesian phylogenetic inference using DNA sequences: A Markov Chain Monte Carlo Method. *Molecular Biology and Evolution, 14*(7), 717–724.

Yang, Z., & Rannala, B. (2006). Bayesian estimation of species diver-gence times under a molecular clock using multiple fossil calibrations with soft bounds. *Molecular Biology and Evolution, 23*(1), 212–226. doi: https://doi.org/10.1093/molbev/msj024

Yang, Z., & Rannala, B. (2012). Molecular phylogenetics: Principles and practice. *Nature Reviews Genetics, 13*(5), 303–314.

Yoder, A. D., & Yang, Z. (2000). Estimation of primate speciation dates using local molecular clocks. *Molecular Biology and Evolution, 17*(7), 1081–1090.

Yule, G. (1925). A mathematical theory of evolution, based on the conclusions of Dr. J. C. Willis, FRS. *Philosophical Transactions of the Royal Society of London. Series B, 213*, 21–87.

Zhang, C., Stadler, T., Klopfstein, S., Heath, T. A., & Ronquist, F. (2016). Total-evidence dating under the fossilized birth-death process. *Systematic Biology, 65*(2), 228–249. doi: https://doi.org/10.1093/sysbio/syv080

Zuckerkandl, E., & Pauling, L. (1962). Molecular disease, evolution, and genetic heterogeneity. In M. Kasha & B. Pullman (Eds.), *Horizons in biochemistry* (pp. 189–225). New York: Academic Press.

Zuckerkandl, E., & Pauling, L. (1965). Evolutionary divergence and conver-gence in proteins. *Evolving Genes and Proteins, 97*, 97–166.

Zwickl, D. J. (2006). Genetic algorithm approaches for the phylogenetic anal-ysis of large biological sequence datasets under the maximum likelihood criterion. Unpublished doctoral dissertation, The University of Texas at Austin.

Acknowledgments

We thank Sandra Álvarez-Carretero, David Bapst, Joëlle Barido-Sottani, Jeremy Brown, Michael Landis, Lee Hsiang Liow, and one anonymous reviewer for feedback that helped greatly improve the manuscript. R. C. M. W. was funded by the ETH Zürich Postdoctoral Fellowship and the Marie Curie Actions for People COFUND program. A. M. W. was funded by a grant from the Louisiana Biomedical Research Network, a National Institutes of Health Institutional Development Award.

Cambridge Elements ⹀

Elements of Paleontology

Editor-in-Chief

Colin D. Sumrall
University of Tennessee

About the Series

The Elements of Paleontology series is a publishing collaboration between the Paleontological Society and Cambridge University Press. The series covers the full spectrum of topics in paleontology and paleobiology, and related topics in the Earth and life sciences of interest to students and researchers of paleontology.

The Paleontological Society is an international nonprofit organization devoted exclusively to the science of paleontology: invertebrate and vertebrate paleontology, micropaleontology, and paleobotany. The Society's mission is to advance the study of the fossil record through scientific research, education, and advocacy. Its vision is to be a leading global advocate for understanding life's history and evolution. The Society has several membership categories, including regular, amateur/avocational, student, and retired. Members, representing some 40 countries, include professional paleontologists, academicians, science editors, Earth science teachers, museum specialists, undergraduate and graduate students, postdoctoral scholars, and amateur/avocational paleontologists.

Paleontological
S O C I E T Y

Cambridge Elements $\equiv$

Elements of Paleontology